FLORA OF TROPICAL EAST AFRICA

TURNERACEAE

J. Lewis

Herbs, shrubs or trees, usually pubescent, sometimes stellately. Leaves alternate, simple, often glandular ; stipules small or absent. Flowers solitary to numerous, axillary and sometimes terminal, in racemes, panicles or cymes, regular, hermaphrodite, sometimes heterostylous. Calyx 5-merous, connate at least near the base, lobes imbricate in aestivation (2 wholly inside, 2 wholly outside). Petals (contorted in aestivation) and stamens 5, both adnate to the calyx-tube forming an hypanthium (see Fig. 1, p. 3) which may be very short (0·25 mm.). Filaments often flattened or narrowly winged ; anthers introrse, dorsifixed, 2-celled, dehiscing longitudinally. Ovary superior, unilocular ; placentae usually parietal and pluriovulate, rarely basal and uniovulate (*Stapfiella* only) ; ovules anatropous. Styles 3 ; stigmas apically divided (at least in our species), usually fimbriate or laciniate. Fruit a 3-valved capsule. Seeds arillate ; aril dry (in our species a unilateral scarious membrane) ; testa hard, either with a raised rectangular network, the spaces of which form longitudinal lines of contiguous pits, or longitudinally striate-submuricate (*Stapfiella* only).

A tropical and subtropical family with very characteristic seeds. The cultivated varieties of *Turnera ulmifolia* L. are not known from our area but may well flourish on introduction ; a satisfactory rockery plant might be developed from *Wormskioldia brevicaulis* var. *rosulata*.

Shrubs, up to 3 m. (not less than about 1 m.) tall :
  Flowers solitary ; capsules 7–10 mm. long, several-
    seeded ; seeds curved ; petals more than 1·5 cm.
    long ; small shrub, about 1 m. tall .   .   . 1. **Loewia**
  Flowers paniculate ; capsules less than 5 mm. long,
    1-seeded ; seed straight ; petals not more than
    1 cm. long ; shrub, about 3 m. tall .   . 2. **Stapfiella**
Herbs, usually up to 0·4 m. (not more than 0·5 m.) tall :
  Flowers solitary, concealed in axils of apically crowded
    leaves ; capsules up to 2·5 mm. long; stem
    strigose ; low plants, up to 20 cm. tall .   . 3. **Hyalocalyx**
  Flowers usually several and obvious ; leaves ± uni-
    formly dispersed or rosetted ; capsules more
    than 3 mm. long ; plants stemless or stems not
    strigose ; mature plants usually more than
    20 cm. tall :
    Seeds straight ; capsule at least 3½ times as long as
      broad, often 20 times or more, not dehiscing
      from the apex ; petals ligulate opposite inser-
      tion .   .   .   . 4. **Wormskioldia**
    Seeds curved ; capsule not more than 3 times as
      long as broad, often less, dehiscing from apex ;
      petals eligulate   .   .   .   . 5. **Streptopetalum**

1. LOEWIA

Urb. in Ann. R. Ist. Bot. Rom. 6 : 189 (1897)

Shrubs ; densely hirsute or pubescent, often including stellate hairs ; glandular. Stipules absent (? or obscure). Bracts leaflike. Flowers solitary in upper leaf axils, erect, trumpet-shaped, bibracteolate, heterostylous (or probably so[1]). Calyx connate for about half its length or more. Petals white, yellow or bright orange, usually inserted at or near the opening of the calyx-tube.[1] Placentae parietal. Capsule narrowly obovoid, dehiscing loculicidially from apex downwards ; beak very short. Seeds several, curved ; pits 2-pored.

A genus closely allied to *Turnera* L., but confined to E. and NE. Africa

NOTE. Adequately flowering material of this genus is scarce in herbaria possibly because the flowers of its members have a very short mature phase, as do those of the herbaceous genera.

Young stem and leaves densely hirsute, not glandular ;
    leaves singly serrate ; calyx connate for ± half its length   1. *L. thomasii*
Young stem and leaves stellately pubescent and glandular ;
    leaves doubly serrate ; calyx connate for more than two-
    thirds its length .   .   .   .   2. *L. tanaënsis*

1. **L. thomasii** (*Urb.*) *J. Lewis* in K.B. 1953 : 431 (1953). Type : Kenya, Lamu District, Witu, *F. Thomas* 47 (B, holo. †, K, iso. !)

Small shrub ; stem very densely and longly hirsute when young, glabrescent below. Leaves obovate, up to 3 × 2·5 cm., cuneate below into a thick petiole (up to 7 mm. long) with a marginal gland at the junction, upper two-thirds of margin sharply serrate, apex acute, densely ciliate on both surfaces. Pedicel, bracteoles and calyx densely ciliate. Pedicel ± 2 mm. long ; bracteoles elliptic-lanceolate, up to 15 mm. long, margin entire or shortly lobed. *Flowers probably heterostylous.*[2] Calyx-tube 12 mm. long ; lobes very narrowly oblong, 14 mm. long. Petals *white*, ± 30 mm. long. *Filaments adnate for ± 5 mm., 20 mm. or more long. Ovary erect-pilose ; style 28 mm. long.* Capsule erect, ovoid, 7 × 5 mm., subangular, densely hirsute. Seeds 3–4 mm. long ; aril 3–4 mm. long, margin and apex irregularly lobed.

KENYA. Lamu District, Witu, 1896, *F. Thomas* 47 !
DISTR. **K**7 ; not known elsewhere.
HAB. Dry sandy places ; ± 25 m.

SYN. *Turnera ulmifolia* L. var. *thomasii* Urb. in E.B.J. 25, Beibl. 60 : 11 (1898). Type as species.

NOTE. The heavy indumentum should make *L. thomasii* easily recognizable. It may be less rare in the field than it appears, and collectors should try to reduce its rarity in herbaria.

2. **L. tanaënsis** *Urb.* in E.J. 25, Beibl. 60 : 2 (1898) ; Hutch. in Ic. Pl. t. 3015 (1915). Type : Kenya, near R. Tana, Korokoro, *F. Thomas* 48 (B, holo. †, K, iso. !)

Small shrub ; stem simply and stellately pubescent when young, glabrescent below and usually with scattered ? glandular tubercles which blacken on drying. Leaves obovate, up to 6 × 4 cm., attenuating below into a very narrowly and partially winged petiole (not more than 7 mm. long), upper two-thirds of margin ± bluntly and doubly serrate, apex obtuse, ± densely

---

[1] See description of *L. thomasii* and footnote below.
[2] Italics in this description indicate quotations from Urban in E.J. 25, Beibl. 60 : 11 (1898) ; isotype not dissected as it bears only two flowers. Degree of adnation of petals to calyx-tube not known in this species.

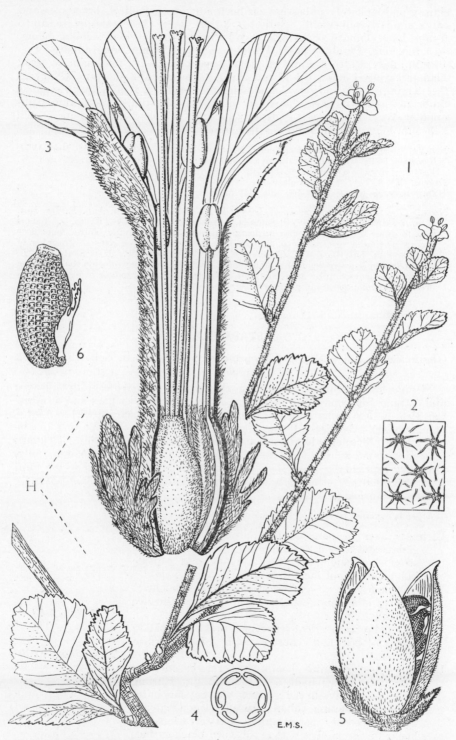

FIG. 1. *LOEWIA TANAENSIS*—1, habit, × 1 ; 2, stem indumentum, × 12 ; 3, bracteoles and partially dissected flower showing hypanthium (H), × 6 ; 4, diagrammatic T.S. ovary ; 5, capsule, × 4 ; 6, seed, × 8. 1–4 & 6, from *Bally* 1998 ; 5, from *Battiscombe* 522.

simply and stellately pubescent on both surfaces. Pedicel, bracteoles and
calyx similarly pubescent ; pedicel ± 3 mm. long ; bracteoles ovate, up to
6 mm. long, pinnatipartite. Calyx-tube up to 15 mm. long ; lobes oblong,
± 5 mm. long. Petals (yellow or) bright orange, ± 24 mm. long, adnate for ±
14 mm., max. width 6 mm. Stamens adnate and hairy for 1 mm. or less ;
filaments winged below, 12–17 mm. long, 3 longer and 2 shorter. Ovary ±
7 × 3 mm., shortly and very densely hirsute ; styles ± 16 mm. long ; stigmas
fimbriate. Capsule erect, narrowly ovoid or obovoid, 10 × 5 mm. subangular, very shortly and ± densely hirsute. Seeds 3–4 mm. long ; aril
2–3 mm. long, margin and apex irregularly lobed. Fig. 1, p. 3.

KENYA. Northern Frontier Province : between Burdale and Kashe, 10 Mar. 1934,
    Sampson 36 ! ; Tana River District ; Garissa, 26 Dec. 1942, *Bally* 1998 *in C.M.*
    12440 ! ; Kericho District : Sotik, Aug. 1931, *Steele* 3979 !
DISTR.   K1, 5, 7 ; not known elsewhere.
HAB.   Riverine forest ; 200–600 and 1830 m.

NOTE.   The third known species of *Loewia* is *L. glutinosa* Urb. in Ann. R. Ist. Bot.
Rom. 6 : 189 (1897). Type : " Somalia," *Brichetti* 3536 (F, holo., K, iso. !). This is
said to differ from *L. tanaënsis* in being glutinous and in having elliptic-lanceolate
bracteoles similar to those of *L. thomasii* ; it does differ in the sizes of a number of
parts and in that the leaves attenuate more gradually below. It is likely that *L.
tanaënsis* is only a variety of this Somaliland species ; a study to decide their relation
requires the collection of more material of both entities, including observations on
bracteole development and glandular secretion in the genus.

2. STAPFIELLA

Gilg in Z.A.E. 571 (1913) in *Flacourtiaceae* ; J. Lewis in K.B. 1953 :   282
(1953)

Medium sized, frequently glandular, shrubs with ± narrowly elliptic leaves
and ± numerous small (less than 1 cm. long) flowers in axillary and terminal
panicles. Calyx connate for less than half its length, bearing within 5 large
(? glandular) tubercles. Petals eligulate, inserted at the opening of the
calyx-tube. Filaments terete, adnate to calyx-tube for ± 0·25 mm. Stigmas
fringed. Ovule solitary and basal. Capsule ± ellipsoid, 4–5 mm. long,
dehiscing septicidally from the apex downwards ; beak less than 0·5 mm.
long. Seed solitary, straight, narrowly obovoid, filling capsule, longitudinally
striate-submuricate.

A genus confined to tropical Africa.

Ultimate axes of panicle only simply pubescent or
    glabrescent ; buds 2–4 mm. long ; fruiting pedicels
    straight or slightly curved ; ovary and capsule
    indumentum not including tubercle-based hairs :
  Stem pubescent ;   leaf-serrations rarely hooked ;
    calyx-lobes and capsule without black surface
    glands . . . .   1. *S. claoxyloides*
  Stem glabrescent ; leaf-serrations mostly hooked ;
    calyx-lobes and capsule with black surface
    glands . . . .   2. *S. ulugurica*
Ultimate axes of panicle bearing tubercle-based hairs
    near the apex ; buds 5–7 mm. long ; fruiting
    pedicels sharply curved ; ovary and capsule indu-
    mentum including tubercle-based hairs (Fig. 2) ;
    stem pubescent ; petals white ; leaf-serrations not
    hooked ; calyx-lobes with ± obscure black surface
    glands and capsule without them   .   .   .   3. *S. usambarica*

FIG. 2. *STAPFIELLA USAMBARICA*, from *Brenan & Greenway* 8299—1, habit, × ½ ; 2, leaf-margin, × 3 ; 3, apex of fertile shoot, × 1½ ; 4, flower with one '' sepal '' and petal removed and ovary sectioned, × 8 ; 5, insertion of petals and stamen on calyx, × 8 ; 6, bunch of tuberculate-based hairs on ovary × 60 ; 7, dehisced capsule, × 8 ; 8, seed, × 8.

1. **S. claoxyloides** *Gilg* in Mildbr., Z.A.E. 571 (1913) ; Staner in Ann. Soc. Sci. Brux., sér. 2, 58 : 106 (1938) ; J. Lewis in K.B. 1953 : 282 (1953). Type : Ruanda Urundi, Rugege forest, Rukarara, *Mildbraed* 920 (B, holo. †) and Belgian Congo, Kabango, *Bequaert* 6174 (BR, neo. !, K, photo !)

Branches rusty brown, ± shortly and thickly grey-pubescent. Leaves 4–8·5 × 1–2·5 cm. ; margin serrate, serrations " gland "-tipped, rarely hooked ; apex acute or subacuminate ; ± uniformly pubescent on both surfaces, especially on the midrib above and on (all) the nerves below ; no gland-dots by transmitted light ; petioles up to 1·5 (–2) cm. long, uniformly pubescent. Some of the bracts within the panicle foliaceous, up to 5 cm. long ; ultimate bracts very small (less than 2 mm. long) ; all axes ± uniformly pubescent. Calyx-tube ± 0·25 mm. long ; tubercles free apically, not fringed, margin sometimes bearing a few long hairs ; calyx-lobes ± 1·75 × 0·5 mm. sparsely pubescent, not gland-dotted. Petals (? greenish-) white, ± 2 mm. long in mature bud. Ovary ± 0·75 × 0·3 mm., densely ciliate. Fruiting pedicels straight or slightly curved ; capsules irregularly pubescent, not black gland-dotted. Aril more than half length of seed.

UGANDA. Kigezi District : north-west, Sept. 1947, *Dale* 496 !
DISTR. U2 ; also in eastern Belgian Congo and Ruanda-Urundi
HAB. (Probably) wooded grassland ; 1200–1500 m.

2. **S. ulugurica** *Mildbr.* in N.B.G.B. 11 : 943 (1933). Type : Tanganyika, Uluguru Mts., Lupanga Peak, *Schlieben* 2973 (B, holo. †, BR, iso. !, K, photo-iso. !)

Branches pale brown, glabrescent, residual indumentum confined to strips. Leaves 4–9·5 × 1–2·5 cm. ; margin serrate, serrations " gland "-tipped and hooked towards the leaf axis ; apex subacuminate ; uniformly pubescent on both surfaces, especially on the midrib towards base above and on (all) the nerves beneath ; uniformly black gland-dotted when dry by transmitted light ; petioles up to 1·5 cm. long, pubescent. Bracts within the panicle usually small, occasionally enlarged up to 2 cm. long and foliaceous ; all axes shortly pubescent or glabrescent. Calyx-tube ± 0·25 mm. long ; tubercles free apically, fringed ; calyx-lobes 1·75 × 0·75 (–1) mm., shortly pubescent, distinctly gland-dotted. Petals yellow or white, ± 2 mm. long in mature bud. Ovary 0·5 × 0·3 mm. densely ciliate. Fruiting pedicels straight or slightly curved. Capsules ± 4·5 × 2 mm., irregularly pubescent, especially above, with a few large distinct black gland-spots. Aril not more than half length of seed.

TANGANYIKA. Morogoro District : Uluguru Mts., Lupanga Peak., 12 Nov. 1932, *Schlieben* 2973 ! and Tanana, Matombo road, Feb. 1935, *E. M. Bruce* 806 !
DISTR. T6 ; endemic on Uluguru Mts.
HAB. Upland rain forest ; 1900–2100 m.

3. **S. usambarica** *J. Lewis* in K.B. 1953 : 282 (1953). Type : Tanganyika, W. Usambara Mts., Magamba Peak, *Brenan & Greenway* 8299 (K, holo. !, EA, IFI, iso. !)

Branches rusty brown, uniformly and very densely grey pubescent. Leaves 4–11 × 1–3 cm. ; margin serrate, serrations occasionally " gland "-tipped, not hooked ; apex acute, rarely sub-acuminate ; midrib impressed and pubescent above, especially towards base, midrib and venation sparsely pubescent beneath, rest of lamina glabrous or subglabrous (on both surfaces) ; uniformly black gland-dotted by transmitted light when dry ; petioles up to 1·5 cm. long, uniformly pubescent. Bracts within the panicle small or foliaceous (up to ± 2 cm. long) ; panicles proximally pubescent only, pubescence distally interspersed with or replaced by ± numerous 1 mm. long multicellular hairs with bases expanded into tubercles, apex of pedicel puberulous (only). Calyx-tube ± 1 (–1·5) mm. long ; tubercles wholly adnate, neither free nor fringed apically ; calyx-lobes 3–4 × 1 mm., puberu lous below, glabrous or subglabrous above, obscurely and sparsely black gland-dotted by transmitted light when dry. Petals white, up to 7 mm. long. Ovary ± 1 × 0·5 mm., ± longly pubescent and tuberculate-hairy. Fruiting pedicels sharply curving-reflexed. Capsules ± 4·25 mm. long, puberulous below, pubescent and tuberculate-hairy above, not black gland-dotted when dry. Aril less than half length of seed. Fig. 2, p. 5.

TANGANYIKA. Lushoto District : Magamba Forest, Oct. 1934, *Pitt-Schenkel*, 408 ! and Shagayu Peak, 24 May 1953, *Drummond & Hemsley*, 2733 !
DISTR. **T3** ; known only from the Western Usambara Mts.
HAB. Upland rain forest [1] ; 1900–2230 m.

3. HYALOCALYX
Rolfe in J.L.S. 21 : 257 (1884)

Hirsute herbs with exstipulate leaves and small (± 4·5 mm. long) solitary axillary heterostylous flowers. Calyx hyaline, connate for about half its length, appendaged. Petals and stamens adnate to calyx very near the base of the calyx-tube. Filaments flattened below ; anthers very broadly oblong. Placentae parietal, each about 4-ovulate ; stigmas laciniate. Capsule dehiscing loculicidally from apex, beak absent. Seeds several, curved. Aril not less than half length of seed.

A monotypic E. and SE. African genus.

H. setifer *Rolfe* in J.L.S. 21 : 258, tab. 7 (1884) as *H. setiferus* ; Perrier de la Bâthie in Arch. Bot., Bull. Mens. 4 : 7 (1930) as *H. setiferus* ; Drummond & Meikle in K.B. 1950 : 335 (1951). Type : NW. Madagascar, Nossibé island, *Rutenberg* (BREM, holo., K, iso. !)

Annual up to 0·2 m. high. Stem brown, woody below, strigose. Leaves bunched terminally, elliptic-oblanceolate, 1·5–4 × 0·5–1 cm., longly hirsute, cuneate below ; margin serrate above ; apex acute ; petiole (1–) 2–5 mm. long. Flowers solitary, ± 4·5 mm. long, concealed in the axils of bracts ; pedicel short, longly hirsute ; bracteoles absent. Calyx ± 2 mm. long, lobes oblong, rounded apically, with 1–3 sub-apical hairlike appendages. Petals yellow, ± 4·25 mm. long. Ovary ± narrowly ovoid, subglabrous. Fruiting peduncles longly accrescent and recurving, ascending-hirsute ; capsule becoming inverted, rotund, 2–2·5 × 1·5 mm., margins of loculi puberulous. Seeds ± 1–1·5 mm. long, several, curved. Aril rotund, ± 1 mm. long, margin entire. Fig. 3, p. 8.

TANGANYIKA. Lindi District : 150 km. W. of Lindi, near Lukuledi, 20 Apr. 1935, *Schlieben* 6338 !
DISTR. **T8** ; Portuguese East Africa and Madagascar.
HAB. Locally common along roadsides in open woodlands and native cultivations ; about 340 m.
SYN. *Turnera setifer* (Rolfe) Baill. in Bull. Soc. Linn. Paris 1 : 582 (1886)
    H. dalleizetti Cap. in Bull. Herb. Boiss., ser 2, 8 : 252 (1908). Type : Madagascar, near Tananarive, *d'Alleizette* (P, holo. !, K, photo. !)

BIOLOGICAL NOTES ON THE FOLLOWING GENERA

The germination in the two genera *Wormskioldia* and *Streptopetalum* is epigeal and consequently heterophyllous members may bear three types of functionally leafy organs. The heterostyly is worthy of field study ; examples of the unusual mixed style-length form recorded by Urban of *S. serratum* should be sought for.

The flowering period may be extended by the production of a second crop of flowers, even (*W. longepedunculata*) extending beyond the end of the rainy season. The individual flowers are relatively transient, expanding from the mature bud during the morning of a sunny day and fading the same evening. This phenomenon makes the estimation of lengths within the flower inconclusive in herbarium material ; where single figures are given below they are the maxima which have been observed.

Variation in number of floral parts has been rarely observed : 4 calyx-lobes (A. Richard) ; a single 6-stamened flower and single 4-styled and 5-styled flowers.

[1] See Pitt-Schenkel, J. Ecol. 26 : 57 (1938).

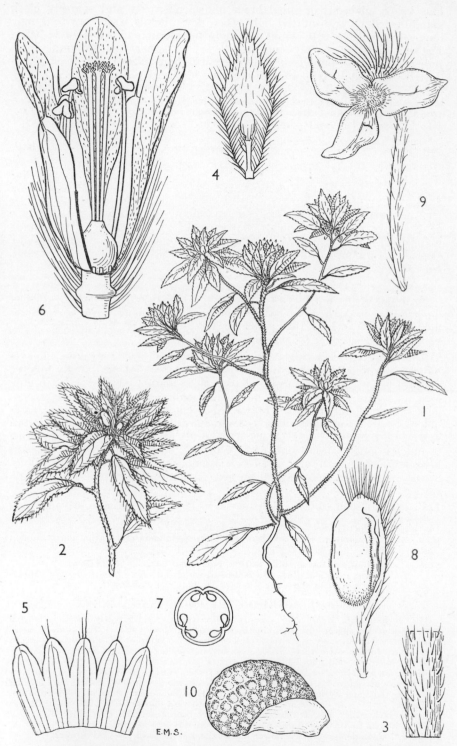

E.M.S.

FIG. 3. *HYALOCALYX SETIFER*, from *Faulkner* 234—1, habit, × ⅔ ; 2, branch apex, × 1½ ; 3, segment of stem, × 8 ; 4, bract and bud, × 4 ; 5, opened calyx, × 8 ; 6, partially dissected flower, × 16 ; 7, diagrammatic T.S. ovary ; 8, capsule, × 8 ; 9, dehisced capsule, × 8 ; 10, seed, × 32.

4. WORMSKIOLDIA

Thonn. in Schum. & Thonn., Beskr. Guin. Pl. 165 (1827)

Annual and perennial herbs, pubescent or puberulous and setiferous. Leaves sometimes glandular ; stipules absent. Inflorescences axillary or scapose, one-sided racemes ; bracteoles single or paired ; pedicels usually accrescent and sometimes curving. Flowers erect, heterostylous (? in annuals only) ; hypanthium 0·3–1·5 mm. long. Calyx-tube hairy within for 2–5 mm. from base, usually bearing internally 5 large (? glandular) tubercles above the insertion of the stamens (lacking only in *W. pilosa*). Petals yellow to scarlet, rarely white, adnate to calyx for about half length of the tube or less, broadly oblanceolate or spathulate, bearing a ligule opposite insertion. Filaments often winged, adnate to calyx for 2·0 mm. or less (but see note on *Streptopetalum graminifolium* ; p. 18). Capsules linear or narrowly ellipsoid, sometimes held horizontally or reflexed ; dehiscence loculicidal, irregular (not from apex downwards). Seeds numerous, straight ; pits 2-pored.

An exclusively tropical and South African genus ; widespread as weeds of cultivation south of latitude 15° N. See Biological Notes, p. 7.

VARIATION. In several species of *Wormskioldia* there can be observed the curious phenomenon that relatively narrower-leaved and more entire-leaved forms occur in the east central part of the genus' range and that relatively broader-leaved and more divided-leaved forms occupy the southern and/or sometimes the northern localities. These morphological variations have in some cases the appearance of clines, possibly induced by climatic differences ; collectors should note carefully the prevailing phase of the seasonal cycle.

Annuals ; caulescent ; flowers 1–4 per inflorescence ; petals 15–20 mm. long :
  All leaves ovate ; stem with straight spreading pubescence and setae ; fruiting pedicels straight . . . . 1. *W. biviniana*
  Heterophyllous ; upper leaves lanceolate ; stem crisped-adpressed-puberulous ; fruiting pedicels curved :
    Margins of upper leaves very usually bluntly lobed below ; stem-setae ± sparse, appearing as ± dark (when dry), yellowish, inconspicuous (± 0·5 mm. long) protuberances 2. *W. lobata*
    Margins of upper leaves sharply lobed or partite, at least below ; stem-setae obvious, at least at upper-nodes, purplish (when dry), 1–4 mm. long . . . . . 3. *W. pilosa*
Perennials ; caulescent or acaulescent ; flowers (1–) 2–14 per inflorescence ; petals 25–40 mm. long :
  Peduncles ± spreading pubescent, sometimes glabrescent ; stem and/or peduncles or pedicels setiferous ; setae yellow or drying black, bulbous-based, 0·5–1·5 mm. long :
    Capsules upright, longly pubescent, 20–30 times as long as broad ; beak more than 3 mm. long ; leaves longly pubescent beneath . 4. *W. prittwitzii*

Capsules horizontal, not longly pubescent, 5–
    10 times as long as broad ; beak not
    more than 3 mm. long ; leaves shortly
    scabrid beneath .   .   .   .   .   5. *W. brevicaulis*
Peduncles not spreading pubescent ; stem setae
    not bulbous-based, purplish when dry, (1–)
    2–4 mm. long, or rarely absent .   .   .   6. *W. longepedunculata*

1. **W. biviniana** *Tul.* in Ann. Sci. Nat., sér. 5, 9 : 324 (1868) ; Urb. in Jahrb. K. bot. Gart. & Mus. Berl. 2 : 50 (1883) ; E. & P. Pf., ed. 2, 21 : 463 (1925). Type : Zanzibar, *Boivin* (P, holo.!, K, photo.! and tracing!)

Caulescent annuals up to 0·5 m. tall. Stem shining-white-pubescent and setiferous ; setae ± 1 mm. long, bulbous-based, golden with bases blackened (when dried). Leaves all ovate, up to 8·5 × 4 cm., blade longly cuneate into a narrowly winged petiole ; margin bi- (tri-) serrate, apex acute, indumentum as stem but pubescence shorter and setae ± confined to the nerves below. Peduncles not longer than the subtending leaves, indumentum as stem ; bracteoles ± 1 mm. long, pubescent ; pedicels 3–9 mm. long, indumentum as stem ; flowers 1–3 per inflorescence, heterostylous. Calyx indumentum as stem ; calyx-tube ± 4·5 mm. long, longly and densely white-hairy within on lower 2·5–3 mm. ; tubercles elliptic, 2 mm. long ; calyx-lobes 4 mm. long. Petals yellow, ± 15 × 2 mm, adnate for ± 5 mm. ; ligule lanceolate, 1·5 mm. long, apex acute. Stamens adnate for ± 0·4 mm. ; filaments winged, 5–6 mm. long, 2 shorter and 3 longer. Ovary ± 5 × 1 mm., densely pubescent ; styles 3–4 mm. long, stigma deeply laciniate. Mature capsules upright, on straight pedicels, (1·5–) 2–4·5 (–4·8) cm. × 1–1·5 mm., pubescent and ± sparsely setiferous ; beak 3–7 mm. long. Aril half the length of the seed or less, rectangular, apex irregularly crenate or lobed.

TANGANYIKA. Mwanza, *Davis* 172!
ZANZIBAR. Zanzibar Is., near Ziwani, Feb. 1931, *Vaughan* 1873!
DISTR. T1 ; Z ; A.-E. Sudan.
HAB. Known as weeds of cultivation, occurring near sea, Lake Victoria and rivers ; 0–1150 m.

NOTE. This species is related to *W. glandulifera* Klotzsch which has relatively longer peduncles and capsules, the capsules being pubescent only, and longer, narrower and more shortly petiolate leaves. This SE. African species is also found near rivers, but is geographically isolated from our species. See note under *W. prittwitzii*, p. 13.

2. **W. lobata** *Urb.* in Jahrb. K. bot. Gart. & Mus. Berl. 2 : 52 (1883) ; Thonner, Fl. Pl. Afr. tab. 106 (1908) ; E. & P. Pf., ed. 2, 21 : 463 (1925) ; Andrews, Fl. Pl. A.-E. Sudan 1 : 29, fig. 22 (1950). Types : Kenya, near Kitui, *Hildebrandt* 2774 (B, syn. †) ; Angola, near Pungo Andongo, *Welwitsch* 2493 (B, syn.†, K & BM, iso.-syn.!)

Caulescent heterophyllous annuals up to 0·3 m. tall. Stem inconspicuously crisped-puberulous ; setae, when present, sparse, short (not above 0·5 mm. long), bulbous-based, dark yellow to brown. Lower (2–3) leaves elliptic, up to 4·8 × 1·2 cm., including petiole of ± 4 mm. long, margin entire or bluntly serrate ; upper leaves sessile or subsessile, lanceolate, up to 12·5 × 3 cm. ± roughly pubescent, rarely subglabrous above ; margins usually shortly and bluntly 1–3 (–4)-lobed, frequently bearing a pair of small pubescent auriculate glandular lobes basally, otherwise entire or ± shallowly serrate or crenate ; apices acute. Flowers 1–4 per inflorescence ; peduncles usually not shorter than subtending leaves, up to 20 cm. long, longly pubescent below, otherwise indumentum as on stem ; bracteoles pubescent, up to 1 mm. long ; pedicels up to 3 (–4) mm. long. Calyx pubescent and sometimes setiferous ; tube 4 mm. long, hairy within on lower 4·5 mm., tubercles narrowly elliptic, 3 mm. long ; lobes 4·5 mm. long. Petals orange to yellow, rarely white, 17–19 (20) mm. × 2 mm. adnate for 4·5 mm. ; ligule lanceolate, 1·5 mm. long, apex acute. Stamens adnate for 0·3–0·5 mm. ; filaments winged, 10–13 mm. long, 2 shorter and 3 longer. Ovary ± 5 × 1 mm., glabrous ; styles 8–10 mm. long. Fruiting pedicels recurved ; mature capsules at right angle to peduncle, glabrous, up to 5 (–6) mm. long, 1–1·5 mm. diameter ; beak 2–5 mm. long. Aril rectangular, more than half the length of the seed ; apex crenate to lobed.

UGANDA. Acholi District : Omeya Angima, Apr. 1943, *Purseglove* 1513 ! ; Busoga District : Dagusi Is., 16 Jan. 1953, *Wood* 604 !
KENYA. Near Kitui, *Hildebrandt* 2774 !
TANGANYIKA. Shinyanga District : near Mantini Hills, 26 Mar. 1932, *B. D. Burtt* 3735 !; Mpanda District : Ikuu, 20 Jan. 1950, *Bullock* 2282 ! ; Ufipa District : Milepa, 26 Dec. 1946, *Pielou* 46 !
DISTR. U1, 3 ; K4 ; T1, 4, 5 ; northern Belgian Congo, Gabon, southern A.-E. Sudan, Northern and Southern Rhodesia, Nyasaland and Angola.
HAB. Locally common in open situations in bushland among short grass on rocky outcrops, ironstone and sandy soils ; 900–1270 m.

3. **W. pilosa** (*Willd.*) [*Schweinf.* ex] *Urb.* in Jahrb. K. bot. Gart. & Mus. Berl. 2 : 54 (1883) ; V.E. 3 (2) : 593 (1921) ; E. & P. Pf., ed. 2, 21 : 463 (1925) ; F.W.T.A. 1 : 81 (1927) ; Andrews, Fl. Pl. A.-E. Sudan 1 : 30, fig. 23 (1950). Type : Gold Coast, *Thonning* (C, holo. !, K, photo. !)

Caulescent heterophyllous annuals up to 0·4 m. tall. Stem white-crisped-puberulous and setiferous, at least at upper nodes ; setae 1–4 mm. long, with a broad but not bulbous base, ± dark purple when dry, often with a sharp yellow apex. Lower 2–4 (–10) leaves elliptic, up to 6·5 × 2·6 cm., pinnati-partite, apex acute ; upper leaves lanceolate, up to 17 × 6 cm., sharply pinnatifid to pinnati-partite below, margin otherwise entire or serrate, apex acute to acuminate ; all leaves cuneate below, sessile or petiolate (petiole not above 4 mm. long), midrib above ± puberulous and below ± setiferous, lamina otherwise glabrous above and papillose below. Flowers 2–4 per inflorescence ; peduncles not usually shorter than subtending leaves, 3–10 cm. long, ± puberulous and weakly setiferous ; bracteoles glabrous, 2–6 mm. long ; pedicels ± 4 mm. long, subglabrous or setiferous above. Calyx shortly setiferous ; tube 6·5–7·5 mm., sparsely hairy within on lower 3–3·5 mm., no tubercles above stamen-bases ; lobes 3–4 mm. long Petals yellow or orange, 15–18 mm. × ± 2 mm., adnate for 2–3·25 mm. ; ligule narrowly oblong, 0·5 mm. long, apex truncate. Stamens adnate for ± 0·3 mm., filaments winged in lower 5 mm., 6·5–11·5 mm. long, 2 shorter, 3 longer. Ovary 5–7 mm. × ± 1 mm., puberulous ; styles ± 0·5 mm. long. Fruiting pedicels recurved ; mature capsules ± at right angles to peduncle, ± shortly setiferous, up to 7·3 cm. × 1–1·5 mm. ; beak 4–7 mm. long. Aril oblong-ovate, about half the length of the seed, apex entire or irregularly crenate.

UGANDA. Karamoja District : near Lomala Hill, June 1930, *Liebenberg* 199 ! ; Bunyoro District : Kitoba, Apr. 1942, *Purseglove* 1229 ! ; Teso District : Serere, Apr. 1932, *Chandler* 602 !
KENYA. Kisumu, 2 Apr. 1922, *Butler* 55 !
TANGANYIKA. Kondoa District : Kikori escarpment, 21 Febr. 1930, *B. D. Burtt* 2741A !
DISTR. U1–3 ; K5 ; T5 ; generally distributed in West Africa from Senegal to northern Nigeria, also in French Cameroons, Shari, A.-E. Sudan and Portuguese East Africa.
HAB. Rocky places, roadsides and cultivated land. Very common on black soil of *Acacia formicarum* woods in Kondoa District (*B. D. Burtt*) ; 1050–1350 m.

SYN. *Raphanus pilosus* Willd., Sp. Pl. 3 : 562 (1801) ; Pers., Syn. 2 : 209 (1806) *Cleome raphanoides* DC. Prodr. 1 : 240 (1824), *nom illegit.*
     W. heterophylla Schum. & Thonn., Beskr. Guin. Pl. 165 (1827) ; A. Rich. in Guill., Perr. & Rich., Fl. Seneg. Tent. [322] (1831), correction for *W. diversifolia*, see below ; F.T.A. 2 : 502 (1871). Type : Gold Coast, *Thonning* (C, holo. !, K, photo. !)

FIG. 4. *WORMSKIOLDIA BREVICAULIS* var. *BREVICAULIS*—1, habit, × ⅔ ; 2, leaf, × ⅔ ; 3, 4, leaves from other plants to show variation, × ⅔ ; 5, dehisced capsule, × 2 ; 6, seed, × 8. *W. BREVICAULIS* var. *ROSULATA*— 7, habit, × ⅔ ; 8, leaf, × ⅔.] 1, 2, from *Smith* ; 3–6, from *Richards* 1181 ; 7, 8, from *Eggeling* 6378.

W. *diversifolia* A. Rich. in Guill., Perr. & Rich., Fl. Seneg. Tent. 36 (1831) ;
Walp., Rep. 5 : 782 (1845–46), *nom. illegit.*
W. *pilosa* (Willd.) Urb. var. *angustifolia* Urb. in Jahrb. K. bot. Gart. & Mus.
Berl. 2 : 54 (1883). Types : Gold Coast, *Thonning* (C, syn. !, K, photo. !)
W. *pilosa* (Willd.) Urb. var. *latifolia* Urb., *loc. cit.* Types : A.-E. Sudan, *Schwein-furth* 2004 and 1700 (B, syn. †, K, iso-syn. !)

VARIATION. Urban's varieties are no more than extremes of continuous variation in
leaf-width and margin-division. The broad-leaved forms occur to the north and
south (*Stocks*, P.E.A.) of the eastern part of the range (see Variation note to genus,
p. 9)

4. **W. prittwitzii** *Urb.* in F.R. 13 : 152 (1914) ; V.E. 3 (2) : 593 (1921) ;
K.B. 1953 : 284 (1953). Types : Tanganyika, Nyambwa [? Kilimatindi]
Konta Mt., *von Prittwitz* 165 (B, holo. †) and Kondoa District, hills SW. of
Kolo, *B. D. Burtt* 1191A (K, neo. !)

Caulescent perennials to 0·3 m. tall. Stem white-spreading-pubescent and
setiferous, especially above ; setae yellow, bulbous-based. Leaves subsessile,
narrowly lanceolate, up to 9 × 1 cm., cuneate below ; margin remotely
serrate at least below, usually entire and attenuate above to an apiculate
apex ; longly pubescent below, sparsely so above except on nerves and
margin. Flowers 2–4 per inflorescence ; peduncles usually not less than
subtending in length, 4·5–8 cm. long, pubescent and setiferous ; bracteoles
narrowly ovate, 1–1·5 (–2) mm. long, pubescent ; pedicels up to 1 cm. long,
pubescent and setiferous. Calyx pubescent ; tube ± 5 mm. long, hairy
within on lower 3·5 mm. ; tubercles broadly elliptic, ± 1·5 mm. long ;
lobes ± 1 cm. long. Petals orange, up to 3·2 cm. long, more than 6 mm.
wide, adnate for 3·5 mm. ; ligule lanceolate, ± 1 mm. long (immature), apex
acute. Stamens equal, adnate for ± 1 mm. ; filaments winged, ± 1 cm.
long. Ovary 4·5 × 1 mm., pubescent ; styles ± 3·75 mm. long. Fruiting
pedicels not or slightly curved ; mature capsules upright or nearly so,
3–4·5 cm. × 1–1·5 mm., longly pubescent ; beak 4–5·5 mm. Aril sub-
rectangular, more than half the length of the seed, apex irregularly crenate.

TANGANYIKA. Kondoa District : hills SW. and E. of Kolo, 5 Jan. 1928, *B. D. Burtt*
1191A ! and 1191 !
DISTR. T5 ; not known elsewhere.
HAB. Dry, sandy and stony places in open bushland on hill slopes ; 1430–1500 m.

NOTE. This species is the perennial equivalent of the more northerly W. *biviniana* and
the more southerly W. *glandulifera*. It has narrower and more nearly entire leaves
than either, which morphological relationship is comparable with that found *within*
some other species. See Variation note to the genus (p. 9).
A poor specimen in the Kew Herbarium, *Swynnerton* 422, resembles this species.
It has the larger more numerous flowers of a perennial, and a pubescent and setiferous
stem, the setae being of the bulbous-based type. It bears also a terete capsule but
this is glabrous and the leaves have sharply serrate margins. The plant is probably
from western Tanganyika ; the locality being " between Mariekemi and [Lake]
Tanganyika." More material matching this anomalous specimen might enable a
new taxon to be delimited.

5. **W. brevicaulis** *Urb.* in Jahrb. K. bot. Gart. & Mus. Berl. 2 : 51 (1883) ;
J. Lewis in K.B. 1953 : 285 (1953). Type : Zanzibar, *Grandidier* 28 (P,
holo. !, K, photo. ! and tracing !)

Acaulescent or occasionally caulescent perennials, up to 0·35 m. Stem when
present, pubescent but not setiferous. Leaves cuneate below to a sessile
base ; apex, shape and dimensions variable (see varietal descriptions and
Fig. 4), densely subscabrid above and below when young, becoming glabrescent ; midrib sometimes and nerves occasionally ± longly pubescent.
Flowers 2–7 per inflorescence ; peduncles longer or shorter than subtending
leaves, 3–14 cm. long, glabrescent or ± spreading-white-pubescent and
usually setiferous ; setae yellow, bulbous-based, 0·5–1·5 mm. long ;
bracteoles ovate-lanceolate, 1–4 mm. long, apex subacuminate ; pedicels

spreading-white-pubescent and setiferous, 6–16 mm. long. Calyx pubescent and (sparsely) setiferous; tube 7–10 mm. long, hairy within on lower 4·5 mm.; tubercles elliptic, 2·5 mm. long; lobes 5–7 mm. long. Petals yellow or bright orange, up to 4·0 cm. × ± 1·3 cm., adnate for ± 5 mm.; ligule lanceolate, 1–1·5 mm. long, apex acute. Stamens equal, adnate for 1–1·5 mm., filaments winged, 1–1·5 cm. long. Ovary 4–5 cm. × 1–1·5 mm., puberulous or subglabrous; styles 8–14 mm. long. Fruiting pedicels recurved; mature capsules usually at rightangles to peduncle, usually 3-ridged or 3-angled, glabrous or puberulous, 1·5–3·0 cm. × 3–5 mm.; beak 1–2 mm. long. Aril ± ovate (or rectangular), more than half the length of the seed, apex irregularly crenate or subentire. Fig. 4, p. 12.

DISTR. Zanzibar and east African coast-lands from Kenya to northern Portuguese East Africa, inland as far as Abercorn District, Northern Rhodesia.

NOTE. The above description covers a group of plants in which the variation, especially of the leaf shape, is considerable and discontinuous. Too little material is available to enable a final taxonomic decision to be made; the tentative division adopted here maintains Urban's groups but reduces their status. It may be that biological investigation will show that even these varieties are merely seasonal states or, conversely, as the notes below to some extent indicate, other varieties may be distinguishable. The former alternative would be supported by sufficient field observations to show that the phenomenon mentioned in the Variation note to the genus (p. 9) occurs in this species, and has some such explanation as the climatic hypothesis suggested there.

var. **brevicaulis**

Acaulescent or occasionally caulescent perennials up to 0·35 m. tall. Leaves ± very narrowly elliptic, up to 18·0 × 1·5 cm. or narrowly elliptic, up to 22 × 4 cm.; margin serrate or biserrate; apex acute or sub-acuminate. Inflorescences usually shorter than subtending leaves. Fig. 4/1–6.

KENYA. Kwale District: inland, Aug. 1949, *Jeffery* K642 ! and Vanga, Oct. 1892, *C. S. Smith* !
TANGANYIKA. Tanga District: Tanga, 1932, *Geilinger* 519 ! and N. of Moa, 9 Apr. 1952, *Bally* 8142 ! : Rufiji, 2 Feb. 1931, *Musk* 148 !
DISTR. **K7**; **T3, 6**; **Z**; Northern Rhodesia
HAB. Usually in open grassland on sandy soils in open bushland and woodland; 0–850 m.

SYN. *W. brevicaulis* Urb.; V.E.3 (2): 593 (1921); E. & P. Pf., ed. 2, 21: 463 (1925)

VARIATION. Specimens of this variety from our area, found below 450 m., have yellow flowers and (except *Geilinger* 519) very narrowly elliptic leaves. The Northern Rhodesian specimens (*Richards* 113 and 1181) occur at 850 m., have orange flowers and narrowly elliptic leaves (see Fig. 4/1–6)

var. **rosulata** (*Urb.*) *J. Lewis* in K.B. 1953: 285 (1953). Type: Tanganyika, Ulanga District, Mbarangandu, *Busse* 680 (B, syn. †, EA, iso-syn. !, K, photo. !)

Acaulescent perennial up to 0·2 m. tall. Leaves narrowly to broadly elliptic or narrowly ovate, up to 13 × 5·5 cm.; margin crenate-serrate; apex obtuse or (more rarely) acute. Inflorescence longer than subtending leaves. Fig. 4/7–8.

KENYA. Probably Tana River District: Kurawa plain, 31 May 1913, *Werner* !
TANGANYIKA. Mpwapwa District: W. of Ntambi, Chiagaia hills, 17 Feb. 1933, *B.D. Burtt* 4529 !; Songea District: near Peramiho, Nov. 1951, *Eggeling* 6378 !
DISTR. **K7**; **T5, 6, 8**; Nyasaland, Portuguese East Africa
HAB. In open woodland and besides roads and tracks through bushland; 50–1350 m.
SYN. *W. rosulata* Urb. in N.B.G.B. 4: 173 (1905)

VARIATION. The *Werner* specimen cited and one collected by *L. Scott* on 25 Oct. 1887 in Nyasaland (both in Kew Herb.) have all their leaf-apices acute. They are however young or stunted specimens. Otherwise in this variety acute apices only occur rarely on plants which have most of their leaf-apices clearly obtuse.

6. **W. longepedunculata** *Mast.* in F.T.A. 2: 502 (1871); Urb. in Jahrb. K. bot. Gart. & Mus. Berl. 2: 53 (1883), as *W. longipedunculata* (followed by subsequent authors); V.E. 3, 2: 593 (1921); E. & P. Pf., ed. 2, 21: 463 (1925); Burtt-Davy, Man. Fl. Pl. Transv. 1: 119 and fig. 11A (1926). Type: Nyasaland, Manganja Hills, *Meller* (K, holo. !)

Acaulescent or caulescent perennial up to 0·75 m. tall. Stem ± shortly white-pubescent and usually setiferous, especially above ; setae long, 1–4 mm., base not bulbous, ± dark purple when dry, often with a sharp yellow apex. Leaves sessile or subsessile, narrowly lanceolate, ± narrowly elliptic or linear, up to 20 cm. long and 0·2–2 cm. wide, base cuneate, margin entire or serrate, usually also ± pinnatifid to pinnatilobed at least below, apex acute or acuminate, pubescent when young, especially beneath, usually glabrescent or glabrous later, especially above and often also sparsely and shortly setiferous beneath, especially on veins. Flowers 6–12 (–14) per inflorescence ; peduncles longer than subtending leaves, (8–) 14–34 cm. long, glabrous, obscurely puberulous or pubescent, especially below, and/or shortly and sparsely setiferous ; bracteoles lanceolate, up to 8 mm. long, shortly pubescent or glabrescent ; pedicels up to 8 mm. long, shortly setiferous above, accrescent. Calyx setiferous and sparsely puberulous ; tube ± 8·5 mm. long, hairy within on lower 3 mm., tubercles very narrowly elliptic, ± 2 mm. long ; lobes 5·5 mm. long. Petals orange-yellow, vermilion or scarlet, 25–30 (–38) mm. × 5–10 mm., adnate for 3 mm.; ligule lanceolate, 1·5 mm. long, apex acute. Stamens adnate for ± 0·5 mm. ; filaments winged below 7·5–13 mm. long, equal. Ovary ± 5 × 1 mm., uniformly pubescent ; styles ± 10 mm. long. Fruiting pedicels reflexed ; mature capsules inverted, sparsely and ± shortly sub-setiferous or glabrescent, 4–8 cm. × 1–1·5 mm. ; beak 4–9 mm. Aril oblong, more than half the length of the seed ; apex irregularly crenate.

Tanganyika. Tanga District, Duga, June 1893, *Holst* 3166! ; Ulanga District, N. of Mahenge, Mbangala, 20 Feb. 1932, *Schlieben* 1792! ; Songea District : Matagoro, Nov. 1951, *Eggeling* 6364!

Distr. **T3**, 6–8 ; Northern and Southern Rhodesia, Nyasaland, Portuguese East Africa, Transvaal and Bechuanaland

Hab. In *Brachystegia-Isoberlinia* woodland and in scattered-tree grassland, also on roadsides and cultivated land ; (? 600–) 1200–1700 m.

Syn. *W. longipedunculata* var. *integrifolia* Urb. in E.J. 15 : 160 (1893). Type : Nyasaland, near Blantyre, *Last* (B, holo. †, K, iso. !)
      W. longipedunculata var. *bussei* Urb. in N.B.G.B. 4 : 174 (1905). Type : Tanganyika, Kilwa District, near Kilwa-Kivinji, *Busse* 444 (B, holo. †, EA, iso. !)

Note. Poisonous to cattle, but having a repellent odour.

Variation. *W. longepedunculata* is a very variable species, and the specimens segregated as distinct elements by Urban are extremes of continuous variation in leaf shape only. The considerable variation in the indumentum, especially of the leaves, is somewhat discontinuous and some grey-pubescent plants (e.g. *Benson* 1011 from Nyasaland) appear distinct. An especially narrow-leaved form (leaves up to 45 × longer than broad) is *Stocks* 89 (K!) from Portuguese East Africa. The increase in the width and division of the leaves in the southern part of the range is an example of the phenomenon mentioned in the Variation note to the genus (p. 9). The two predominantly southern African species, *W. schinzii* Urb. in E.J. 15 : (1892) ; Type : Mozambique, Gambos, *Newton* 26 (Z, holo. !) and *W. juttae* Dinter & Urb. in F.R. 13 : 153 (1914) ; Type : S.W. Africa, NE. Damaraland, *Dinter* 652 (B, syn. †, B, iso-syn.[1]) could be considered variants of *W. longepedunculata* but, as no fruiting material of the former is available and only one specimen of the latter, their status cannot be altered at present, and the description above excludes them. They differ from our species in having the lobes of the pinnatilobed leaves serrate or pinnatifid and the abaxial leaf surface densely setiferous.

5. **STREPTOPETALUM**

Hochst. in Flora 24 : 665 (1841)

Annual and perennial herbs ; pubescent and setiferous. Leaves without glands or stipules. Inflorescences axillary, 1- to many-flowered, usually in a one-sided raceme ; bracteoles single or obscure ; pedicels sometimes

[1] The specimen accepted as authentic for this species was one named in Dinter's handwriting, and which has been compared with the iso-syntype by Dr. J. Mildbraed. It is an unlocalized specimen without a collector's label, and is in the Berlin Herbarium.

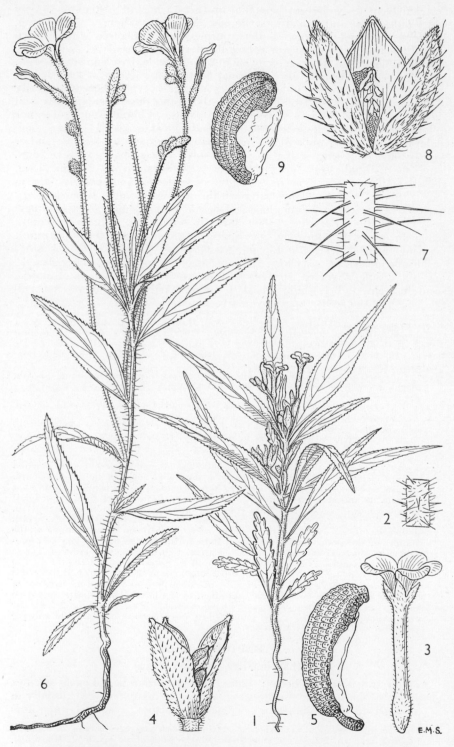

FIG. 5. *STREPTOPETALUM SERRATUM*, from *Bally* 4531—1, habit, × ⅔ ;   2, segment of stem, × 4 ; 3, flower, × 4 ; 4, dehisced capsule, × 4 ; 5, seed, × 12. *S. HILDEBRANDTII*, from *Bally* 7663—6, habit, × ⅔ ; 7, segment of stem, × 4 ; 8, dehisced capsule, × 4 ; 9, seed, × 12.

accrescent, not curving. Flowers erect, sometimes heterostylous, hypanthium ± 3 mm. long. Calyx tube hairy within for 3–6 mm. from base, bearing internally 5 large (? glandular) narrowly elliptic tubercles above the insertion of the stamens. Petals yellow or orange, broadly oblanceolate or spathulate, adnate to calyx for more than half the length of the tube (in our species inserted within 1 mm. of opening of the tube, though not known in *S. graminifolium*), eligulate. Filaments winged, at least in lower third, adnate to calyx for more than 2 mm. (except *S. graminifolium*). Capsules broadly ellipsoid, sub-ovoid or suborbicular, erect, dehiscing loculicidally from the apex. Seeds numerous, curved ; pits 2-pored.

An exclusively tropical and South African genus ; the distribution is similar to that of *Wormskioldia*, but not extending into West Africa. See Biological Notes, p. 7.

Annuals ; stem-setae bulbous-based, yellow or drying
　　black ; capsule shortly and regularly setiferous :
　　Heterophyllous ; leaves 5 mm. or more wide, not
　　　　entire (Fig. 5/1–5)　 .　　.　　.　　.　　. 　 1. *S. serratum*
　　Homophyllous ; leaves 3 mm. or less wide, entire　 2. *S. graminifolium*
Perennial ; stem-setae not bulbous-based, purple
　　and ± yellow-tipped, at least when dry ; capsule
　　pubescent and irregularly setiferous ; homophyl-
　　lous (Fig. 5/6–9)　 .　　.　　.　　.　　.　　. 　 3. *S. hildebrandtii*

1. **S. serratum** *Hochst.* in Flora 24 : 666 (1841) ; Urb. in Jahrb. K. bot. Gart. & Mus. Berl. 2 : 56 (1883) ; V.E. 3 (2) : 594 (1921) ; E. & P. Pf., ed. 2, 21 : 463 (1925) ; W.F.K., 6 (1948). Type : Ethiopia, Tsellenti, Takkaze R., *Schimper* 1260 (B, holo. †, K & BM, iso. !)

Caulescent heterophyllous annual up to 0·4 m. tall. Stem shortly pubescent and ± densely setiferous ; setae short (up to 1 mm. long), yellow, and swollen at base. Leaves ± narrowly elliptic, entire and tapering below into a short (less than 5 mm. long) petiole, glabrous or ± pubescent on both surfaces and shortly setiferous on the midrib and sometimes on the nerves, below ; lower leaves up to 6 × 1 cm., margin symmetrically and bluntly pinnatifid or pinnatilobed, apex acute ; upper leaves up to 12 × 2 cm., margin above bluntly serrate, apex subacuminate. Flowers 1–9 per inflorescence ; peduncle pubescent and setiferous, up to 1·5 cm. long ; bracteoles obscure ; pedicels pubescent and setiferous, ± 1 mm. long, not accrescent. Calyx shortly pubescent and ± sparsely setiferous, tube 8–11 mm. long, hairy within on lower 3–4 mm., lobes 2–4 mm. long. Petals yellow or orange, 13–16 × 2–4 mm. Filaments equal (or two shorter—*Urban*), 9–11 (–12) mm. long. Ovary oblong, 2 × 1 mm., evenly and shortly pubescent (or 1 shorter—*Urban*), 8–9 mm. long. Capsules ± broadly ellipsoid, (6–) 7–10 (or more) × 3–4 mm., shortly and regularly setiferous ; setae often black-based ; beak 0·75 mm. long ; dehisced valves with wavy margins when dry. Aril narrowly oblong, as long as the seed or nearly so ; margin irregularly crenate. Fig. 5/1–5.

UGANDA. Karamoja District : Kokumongole, 28 May 1939, *A. S. Thomas* 2849 !
KENYA. Northern Frontier Province : Dandu, 5 May 1952, *Gillett* 13043 !; Baringo
　　District : Kamasia, Maji ya Moto, 16 July 1945, *Bally* 4531 *in CM.* 12337 !
TANGANYIKA. Near Shinyanga, Jan./Feb. 1932, *Bax* 4 !; Mpwapwa District : Kongwa,
　　24 Feb. 1949, *Anderson* 331 !
DISTR. U1 ; K1–3, 7 ; T1, 3, 5 ; A.-E. Sudan, Ethiopia, Northern Rhodesia (SW.
　　border only), Bechuanaland, Transvaal and South West Africa
HAB. On sandy soil among rocks in bushland and thorn-scrub, and in grassland and by
　　roadsides ; 510–1830 m.

Syn. *W. serrata* (Hochst.) [Hochst. ex] Walp., Rep. 5 : 782 (1845–46) ; F.T.A. 2 :
      502 (1871)
      W. abyssinica A. Rich., Tent. Fl. Abyss. 1 : 299 (1847), *nom. illegit.* Type: as
      S. serratum Hochst.

Note. Attention is drawn to the discontinuity of this species' distribution ; the
      material from the two areas is indistinguishable morphologically. The areas are
      separated by a region of higher rainfall (more than 40 in. per annum) which may
      be the significant factor.

2. **S. graminifolium** *Urb.* in N.B.G.B. 1 : 31 (1895). Type : Tanganyika,
Tabora District, Gonda, *Boehm* 260 (B, holo. †)

Caulescent homophyllous annual up to 0·4 m. tall. Stem both spreading-
yellow-pubescent and crisped-puberulous, and setiferous especially near the
upper nodes ; setae yellow and bulbous-based. Leaves subsessile or petiole
up to 1·5 mm. long, linear, up to ± 10 cm. × 2–3 mm., entire and minutely
substipitate-glandular, both surfaces pubescent, not setiferous. Inflorescence
racemose, up to 20 cm. long, up to 18-flowered ; bracteoles 5–8 mm. long ;
pedicels 1–2 mm. long. Calyx longly pubescent and shortly setiferous, tube
± 6 mm. long, hairy within on lower 2 mm. ; lobes ± 4 mm. long. Petals
not seen complete. Filaments adnate to calyx for 0·5 mm., all 7 mm. long.
Styles very short, scarcely 2 mm. long. Ovary ellipsoid, shortly and densely
hirsute. Capsule up to 6 × 5 mm., very shortly beaked, fairly densely
setiferous below. Seeds curved, ± 2·75 mm. long ; aril exceeding half the
length of the seed.

Tanganyika. Tabora District : Gonda (lat. 5° 33′ S., long. 32° 42′ E.), April 1882,
      Boehm 260.
Distr. **T4** ; known only from the above locality
Hab. On seasonally flooded ground in open woodland ; about 1000 m.

Note. The above description, adapted from that of Urban, shows a number of unique
      features. This is the only *Streptopetalum* with the filament-adnation less than 2 mm.
      long, as it is in *Wormskioldia*, and this measurement is therefore suspect. Heterostyly
      is suspected (" Flores dimorphi ? " *Urban*), heterophylly not observed and the plant
      is reported as an annual. The suspicion arises that this represents a perennial species
      in its first year, and that it is in fact a narrow-leaved variant of a species, the more
      common examples of which have been described more recently under the name *S.*
      wittei Staner in De Wild. & Staner, Contrib. Fl. Katanga, Suppl. 4 : 68 (1932). Types :
      Belgian Congo, Kiambi, *de Witte* 237 and *Luxen* 26 and 50 (all B, syn. !, all K, photo. !)
      a species of the Belgian Congo and N. Rhodesia. Re-collection of *S. graminifolium*
      (in Tanganyika) is required to confirm or deny this hypothesis, in connection with
      which the occurrence of similarly very narrow-leaved variants of *W. longepedunculata*
      is noteworthy.

3. **S. hildebrandtii** *Urb.* in Jahrb. K. bot. Gart. & Mus. Berl. 2 : 57 (1883).
Type : Kenya, Kitui, *Hildebrandt* 2728 (B, holo. †, K & BM, iso. !)

Caulescent homophyllous perennial up to 0·4 m. tall. Stem shortly white-
pubescent and setiferous ; setae 1–4 mm. long, often drying purple with a
pale tip. Leaves ± narrowly elliptic, 3–7 (–10) cm. × 0·5–1·5 (–2·5) cm.,
sessile or petiole not more than 1 (–2) mm. long ; base narrowly cuneate ;
margin sharply serrate, apex acute ; roughly pubescent and setiferous.
Flowers 3–5 (–6) per inflorescence ; peduncle 5–14 cm. long, puberulous and
densely setiferous ; setae yellow, swollen-based and up to 1 mm. long and
sometimes some similar to stem-setae ; bracteoles ± 2 mm. long, deciduous ;
pedicels 3–5 (–8) mm. long, indumentum as of peduncle. Calyx pubescent
and sparsely setiferous ; tube 10–12 mm. long, hairy within on lower
4–6 mm. ; lobes 5–8 mm. long. Petals yellow or orange, 22–32 × 7–8 mm.
Filaments equal, 10–12 mm. long. Ovary ovoid-oblong, 2–3·5 × 1·5 mm.
longly and irregularly pubescent ; styles ± 10·5 mm. long. Capsules broadly

ellipsoid, 4–8·5 × 3·5 mm., long pubescent and irregularly setiferous ; beak 0·5 mm. long (see Fig. 5/8). Aril ovate, exceeding half the length of the seed in length ; margin irregularly subcrenate. Fig. 5/6–9, p. 16.

KENYA. Machakos District : Makueni location, Nov. 1949, *J. G. Williams in Bally* 7663 *in CM.* 17249 ! & Chyulu foothills, 15 May 1938, *Bally* 7636 ! ; Teita District : Mwatate, Oct. 1884, *Johnston* !

TANGANYIKA. Mbulu District : Babati, Oct. 1925, *Haarer* 92B ! ; Moshi District (Kenya boundary) : Chala crater, 2 Dec. 1932, *Geilinger* 4237 ! and between Himo and Taveta, 13 Apr. 1952, *Greenway* 8721 !

DISTR. **K**4, 7 ; **T**2 ; not known elsewhere.

HAB. In open grassland and scattered-tree grassland ; 1000–1250 **m.**

INDEX TO TURNERACEAE